ENERGY SECTOR STANDARD OF THE PEOPLE'S REPUBLIC OF CHINA

中华人民共和国能源行业标准

Technical Code for Post-assessment of Environmental Impacts for Hydropower Projects

水电工程环境影响后评价技术规范

NB/T 10140-2019

Chief Development Department: China Renewable Energy Engineering Institute
Approval Department: National Energy Administration of the People's Republic of China
Implementation Date: October 1, 2019

China Water & Power Press
中国水利水电出版社
Beijing 2024

All rights reserved. No part of this publication may be reproduced, stored in a retrieval system, or transmitted in any form or by any means—electronic, mechanical, photocopying, recording or otherwise, without prior written permission of the publisher.

图书在版编目（CIP）数据

水电工程环境影响后评价技术规范：NB/T 10140
-2019 = Technical Code for Post-assessment of
Environmental Impacts for Hydropower Projects（NB/
T 10140-2019）：英文 / 国家能源局发布. -- 北京：
中国水利水电出版社, 2024. 8. -- ISBN 978-7-5226
-2708-3
Ⅰ. X820.3-65
中国国家版本馆CIP数据核字第2024JC3592号

ENERGY SECTOR STANDARD
OF THE PEOPLE'S REPUBLIC OF CHINA
中华人民共和国能源行业标准

Technical Code for Post-assessment
of Environmental Impacts for Hydropower Projects
水电工程环境影响后评价技术规范
NB/T 10140-2019
（英文版）

Issued by National Energy Administration of the People's Republic of China
国家能源局　发布
Translation organized by China Renewable Energy Engineering Institute
水电水利规划设计总院　组织翻译
Published by China Water & Power Press
中国水利水电出版社　出版发行
　　Tel: (+ 86 10) 68545888　68545874
　　sales@mwr.gov.cn
　　Account name: China Water & Power Press
　　Address: No.1, Yuyuantan Nanlu, Haidian District, Beijing 100038, China
　　http://www.waterpub.com.cn
中国水利水电出版社微机排版中心　排版
北京中献拓方科技发展有限公司　印刷
184mm×260mm　16开本　3印张　95千字
2024年8月第1版　2024年8月第1次印刷
Price（定价）：￥495.00

Introduction

This English version is one of China's energy sector standard series in English. Its translation was organized by China Renewable Energy Engineering Institute authorized by National Energy Administration of the People's Republic of China in compliance with relevant procedures and stipulations. This English version was issued by National Energy Administration of the People's Republic of China in Announcement [2023] No. 4 dated May 26, 2023.

This version was translated from the Chinese Standard NB/T 10140-2019, *Technical Code for Post-assessment of Environmental Impacts for Hydropower Projects*, published by China Water & Power Press. The copyright is reserved by National Energy Administration of the People's Republic of China. In the event of any discrepancy in the implementation, the Chinese version shall prevail.

Many thanks go to the staff from the relevant standard development organizations and those who have provided generous assistance in the translation and review process.

For further improvement of the English version, any comments and suggestions are welcome and should be addressed to:

China Renewable Energy Engineering Institute
No. 2 Beixiaojie, Liupukang, Xicheng District, Beijing 100120, China
Website: www.creei.cn

Translating organization:

POWERCHINA Zhongnan Engineering Corporation Limited

Translating staff:

YANG Hong   LIU Xiaofen   LI Qian   CHEN Lei
ZHANG Dejian

Review panel members:

QIE Chunsheng	Senior English Translator
LI Zhongjie	POWERCHINA Northwest Engineering Corporation Limited
YU Weiqi	China Renewable Energy Engineering Institute
ZHU Xiaoying	Zhejiang University

QI Wen	POWERCHINA Beijing Engineering Corporation Limited
WANG Lei	POWERCHINA Huadong Engineering Corporation Limited
WANG Tianye	POWERCHINA Chengdu Engineering Corporation Limited
LI Shisheng	China Renewable Energy Engineering Institute

National Energy Administration of the People's Republic of China

翻译出版说明

本译本为国家能源局委托水电水利规划设计总院按照有关程序和规定，统一组织翻译的能源行业标准英文版系列译本之一。2023年5月26日，国家能源局以2023年第4号公告予以公布。

本译本是根据中国水利水电出版社出版的《水电工程环境影响后评价技术规范》NB/T 10140—2019 翻译的，著作权归国家能源局所有。在使用过程中，如出现异议，以中文版为准。

本译本在翻译和审核过程中，本标准编制单位及编制组有关成员给予了积极协助。

为不断提高本译本的质量，欢迎使用者提出意见和建议，并反馈给水电水利规划设计总院。

地址：北京市西城区六铺炕北小街2号
邮编：100120
网址：www.creei.cn

本译本翻译单位：中国电建集团中南勘测设计研究院有限公司
本译本翻译人员：杨　虹　刘小芬　李　倩　陈　蕾
　　　　　　　　张德见

本译本审核人员：
　　郄春生　英语高级翻译
　　李仲杰　中国电建集团西北勘测设计研究院有限公司
　　喻卫奇　水电水利规划设计总院
　　朱小萤　浙江大学
　　齐　文　中国电建集团北京勘测设计研究院有限公司
　　王　蕾　中国电建集团华东勘测设计研究院有限公司
　　王天野　中国电建集团成都勘测设计研究院有限公司
　　李仕胜　水电水利规划设计总院

国家能源局

Announcement of National Energy Administration of the People's Republic of China
[2019] No. 4

National Energy Administration of the People's Republic of China has approved and issued 297 sector standards such as *Code for Electrical Design of Photovoltaic Power Projects*, including 105 energy standards (NB), 168 electric power standards (DL), and 24 petrochemical standards (NB/SH).

Attachment: Directory of Sector Standards

National Energy Administration of the People's Republic of China

June 4, 2019

Attachment:

Directory of Sector Standards

Serial number	Standard No.	Title	Replaced standard No.	Adopted international standard No.	Approval date	Implementation date
...						
13	NB/T 10140-2019	Technical Code for Post-assessment of Environmental Impacts for Hydropower Projects			2019-06-04	2019-10-01
...						

Foreword

According to the requirements of Document GNKJ [2014] No. 298 issued by National Energy Administration of the People's Republic of China, "Notice on Releasing the Development and Revision Plan of the First Batch of Energy Sector Standards in 2014", and after extensive investigation and research, summarization of the practical experience, consultation of relevant advanced Chinese standards, and wide solicitation of opinions, the drafting group has prepared this code.

The main technical contents of this code include: basic requirements, project review, investigation and assessment of environmental impacts, investigation and assessment of environmental protection measures, survey of public opinions, and conclusions and recommendations.

National Energy Administration of the People's Republic of China is in charge of the administration of this code. China Renewable Energy Engineering Institute has proposed this code and is responsible for its routine management. Energy Sector Standardization Technical Committee on Hydropower Planning, Resettlement and Environmental Protection is responsible for the explanation of specific technical contents. Comments and suggestions in the implementation of this code should be addressed to:

China Renewable Energy Engineering Institute
No. 2 Beixiaojie, Liupukang, Xicheng District, Beijing 100120, China

Chief development organizations:

China Renewable Energy Engineering Institute

POWERCHINA Zhongnan Engineering Corporation Limited

Participating development organizations:

POWERCHINA Huadong Engineering Corporation Limited

POWERCHINA Guiyang Engineering Corporation Limited

POWERCHINA Kunming Engineering Corporation Limited

Hunan Zhongnan Water and Environmental Protection Technology Co., Ltd.

Chief drafting staff:

YANG Wenzheng	XUE Lianfang	QIU Jinsheng	LU Bo
JIANG Hao	CHU Kaifeng	LI Xiang	SHI Jiayue
TAN Shengkui	TANG Zhongbo	YANG Jie	ZHAO Qi

ZHANG Dejian LI Huang DONG Lian LIU Huangcheng
YAN Jianbo ZHAO Kun YU Weiqi XIAO Wu
XIAO Jingguang

Review panel members:

WAN Wengong RUI Jianliang CHEN Guozhu CUI Lei
YANG Hongbin LI Jinghua CHEN Bangfu CHEN Daqing
LIU Qing LI Min DAI Xiangrong ZHANG Rong
JIANG Hong KOU Xiaomei LIU Guihua ZHANG Demin
LEI Shaoping WU Wenping LI Shisheng

Contents

1	**General Provisions**	1
2	**Basic Requirements**	2
3	**Project Review**	4
3.1	General Requirements	4
3.2	River Hydropower Development	4
3.3	Project Construction and Operation	4
3.4	Environmental Rationality Analysis of Project Change	5
3.5	Implementation of Environmental Protection Work	6
4	**Investigation and Assessment of Environmental Impacts**	8
4.1	General Requirements	8
4.2	Hydrological Regime	8
4.3	Water Temperature	10
4.4	Surface Water Environment	12
4.5	Aquatic Ecosystem	13
4.6	Terrestrial Ecosystem	15
4.7	Climate	17
4.8	Socio-economy	18
4.9	Environmental Impacts of Resettlement	19
4.10	Comprehensive Evaluation	20
5	**Investigation and Assessment of Environmental Protection Measures**	21
5.1	General Requirements	21
5.2	Hydrological Regime	21
5.3	Water Temperature	21
5.4	Surface Water Environment	22
5.5	Aquatic Ecosystem	23
5.6	Terrestrial Ecosystem	23
5.7	Environmental Protection of Resettlement	24
5.8	Comprehensive Evaluation	24
6	**Survey of Public Opinions**	25
6.1	General Requirements	25
6.2	Survey Methods and Respondents	25
6.3	Survey Contents	25
6.4	Analysis and Addressing of Public Opinions	26

7	Conclusions and Recommendations	27
Appendix A	Table of Contents for a Report on Post-assessment of Environmental Impacts of Hydropower Projects	28
Appendix B	Indicators for Post-assessment of Environmental Impacts for Hydropower Projects	31
Explanation of Wording in This Code		35
List of Quoted Standards		36

1 General Provisions

1.0.1 This code is formulated with a view to standardizing the principles, content and methods and unifying the technical requirements for the post-assessment of environmental impacts for hydropower projects.

1.0.2 This code is applicable to the investigation, analysis and assessment on the actual environmental impacts of hydropower projects which have been under steady operation for a certain period.

1.0.3 Post-assessment of environmental impacts for a hydropower project shall, according to the general requirements for environmental protection of the river basin, make an in-depth analysis of the long-term, cumulative and regional impacts of the hydropower project and the environmental changing trend, and conduct an evaluation on the environmental quality status and the implementation effectiveness of the environmental protection measures, following the principles of being objective, impartial, scientific, systematic and targeted.

1.0.4 In addition to this code, the post-assessment of environmental impacts for hydropower projects shall comply with other current relevant standards of China.

2 Basic Requirements

2.0.1 Post-assessment of environmental impacts for hydropower projects shall consider the relevance of such elements as hydrologic regime, water temperature, surface water environment and aquatic ecosystem and exploit the available research data to analyze the actual environmental impacts of the hydropower project from the perspectives of the river basin and the project. Evaluation on the effectiveness of environmental protection measures shall put emphasis on the particularity, systematicness and coordination of the measures taken, and put forward comments and recommendations for subsequent improvement of the measures according to the objectives and requirements of environmental management.

2.0.2 The range of post-assessment of environmental impacts for hydropower projects shall be consistent with the assessment range defined in the environmental impact assessment documents. In the case of any project change or if the environmental impact assessment documents fails to fully reflect the actual environmental impacts caused by the project, the assessment range shall be adjusted as appropriate according to the project changes and the actual environmental impacts as well as the field investigation.

2.0.3 Post-assessment of environmental impacts for hydropower projects should be conducted within three to five years after the project being put into operation, or within the period determined according to the environmental impacts of the project and the changing characters of the environmental factors. The work period shall not be less than one full year.

2.0.4 Post-assessment of environmental impacts for hydropower projects shall be carried out in the following procedure:

1. Collect and analyze the relevant results and data, and work out the work program for the post-assessment of environmental impacts according to the project characteristics and the environmental features.

2. Investigate the project construction and operation, the environmental impacts arising from the project operation, and the implementation of the environmental protection measures.

3. Analyze the actual environmental impacts of the project, evaluate the environmental status and the change trend, evaluate the effectiveness of the environmental protection measures and the environmental management status, and put forward comments and recommendations for subsequent improvement of the measures.

4   Prepare the environmental impact post-assessment report based on the investigation and evaluation outcomes. The contents of the environmental impact post-assessment report of hydropower projects shall be in accordance with Appendix A of this code.

2.0.5 Post-assessment of environmental impacts for hydropower projects shall determine the content of investigation and evaluation, select appropriate evaluation indicators, and establish the evaluation indicator system according to the project characteristics and the regional environmental features, considering the environmental impact assessment documents and the environmental management requirements. The evaluation indicators shall be representative and measurable and reflect the actual impacts of the hydropower project. The indicators for the environmental impact post-assessment of a hydropower project shall be selected from Appendix B of this code.

2.0.6 Post-assessment of environmental impacts for hydropower projects should be conducted using quantitative methods such as typical survey, numerical simulation model and measurement verification. If quantitative methods are not applicable, qualitative methods such as analogy may be used.

2.0.7 The collected data shall be analyzed for reliability, representativeness and validity, and be adopted rationally. If the available data cannot satisfy the requirements of the post-assessment, necessary supplementary investigation and monitoring shall be done.

3 Project Review

3.1 General Requirements

3.1.1 The project review shall collect the hydropower planning documents for the river on which the project is located, the environmental impact assessment documents for the planning, the relevant outcomes of the environmental protection researches and the design documents of the completed project, the environmental impact assessment documents, the project completion environmental protection acceptance investigation documents, etc.

3.1.2 The documents and drawings shall be organized and analyzed, and the accuracy shall meet the requirements for the post-assessment of environmental impacts.

3.1.3 Project review shall investigate the project changes and analyze their environmental rationality.

3.2 River Hydropower Development

3.2.1 For the general condition of the river, the status of the natural resource utilization, the state of the ecology and environment and the socioeconomic conditions along the river on which the project is located shall be investigated, with emphases on the environmental function zoning, the ecological function zoning, the water function zoning, and the distribution of the environmentally sensitive objects.

3.2.2 For the river hydropower development, the hydropower plan, development and utilization of the river on which the project is located shall be investigated, with emphases on the position and role of the project in the river hydropower plan, and the construction, operation and dispatching of the important cascades in the river hydropower plan.

3.2.3 For environmental protection of the river hydropower development, execution of the environmental protection works along the river on which the project is located, environmental protection requirements and implementation of the environmental protection measures related to the project shall be investigated.

3.3 Project Construction and Operation

3.3.1 For the general condition of the project, the background, development tasks, installed capacity, reservoir features, general layout, construction layout, resettlement, construction schedule, operation and dispatching and main techno-economic indices shall be investigated.

3.3.2 For the project construction period, the times for the commencement, river closure, initial impoundment, putting-into-operation and completion acceptance as well as general description of the completed works shall be investigated.

3.3.3 For the project operation period, the actual operation and dispatch shall be investigated, with emphases on the variation of the discharge flow in various time scales, the artificial regulation and control process in the flood season, the assurance of ecological flow, etc.

3.3.4 For the project resettlement, relocation of relocatees, production resettlement, construction of special facilities and related environmental protection work shall be investigated.

3.3.5 For the project changes, the content and reasons of the project changes shall be investigated and the technical documents and the approval documents of the project changes shall be collected.

3.4 Environmental Rationality Analysis of Project Change

3.4.1 For the project change, the nature of the change shall be identified, and the compliance of the change procedure and the actual environmental impacts arising from the change shall be analyzed.

3.4.2 The investigation and analysis of environmental rationality of the engineering scheme change shall be as follows:

1 Investigate and analyze the environmental rationality of the changes in the project purpose, with an emphasis on the environmental rationality of the changes in the development tasks for the operation period.

2 Investigate and analyze the environmental rationality of the change in the reservoir features, with an emphasis on the environmental rationality of the change in the reservoir regulation performance in the operation period.

3.4.3 The investigation and analysis on the environmental rationality of the operation scheme change shall be as follows:

1 Investigate and analyze the environmental rationality of the operation scheme change in the operation period, on the basis of the demands of the downstream surface water environment and the aquatic ecosystem.

2 Investigate and analyze the environmental rationality of the change in the water resource utilization in the operation period, on the basis of the regional water resource utilization situation and the water use demands

for the downstream water supply, irrigation, navigation, landscaping, etc.

3 Investigate and analyze the environmental rationality of the change in the ecological discharge and assurance measures, on the basis of the downstream water use demands.

3.4.4 Investigation and analysis on the environmental rationality of the construction scheme change shall be as follows:

1 Investigate and analyze the environmental rationality of the construction area changes, with emphases on the environmental rationality of the changes in the construction plants, construction camps and construction road layout.

2 Investigate and analyze the environmental rationality of the change in the construction material sources on the basis of cut-fill balancing, with emphases on the environmental rationality of the additional spoil areas, quarries and borrow areas.

3.4.5 Investigation and analysis on the environmental rationality of the resettlement change shall be as follows:

1 Investigate and analyze the environmental rationality of the relocation changes, with an emphasis on the environmental rationality of the changes in the important and/or concentrated host areas.

2 Investigate and analyze the environmental rationality of the production resettlement changes, with an emphasis on the environmental rationality of the land development.

3.5 Implementation of Environmental Protection Work

3.5.1 The investigation range of the environmental protection work shall cover the project construction area, the reservoir area and the resettlement area, and the investigation content shall involve the environmental protection work in different periods including the design, construction and operation.

3.5.2 For the environmental protection work in the design period, the following shall be investigated:

1 Implementation of relevant national laws and regulations and departmental administrative rules.

2 Preparation and approval of environmental impact assessment documents.

3 Preparation and approval of environmental protection design

documents and environmental protection cost estimates.

3.5.3 For the environmental protection work in the construction period, the following shall be investigated:

1. Setup and operation of the environmental management department and the environmental supervision organization.

2. Implementation and any changes of the environmental protection measures.

3. Implementation of the environmental protection monitoring plan.

4. Use of environmental protection investment.

5. Completion acceptance of environmental protection works and handling of the outstanding issues.

3.5.4 For the environmental protection work in the operation period, the following shall be investigated:

1. Formulation and implementation of the environmental management system for the project operation, and set-up and operation of the environmental management department for the project operation.

2. Implementation and operation of the environmental protection and monitoring measures for the project operation.

3. Implementation of laws and regulations, plans and management rules related to the project operation.

4. Implementation of other environmental protection plans and measures related to the project.

4 Investigation and Assessment of Environmental Impacts

4.1 General Requirements

4.1.1 For the environmental impact investigation, field investigation shall be carried out according to the river hydropower development, the environmental impact assessment documents and the project operation characteristics, considerating the environmental conditions of the investigated area and implementation of the environmental protection measures. If the investigation results cannot meet the post-assessment requirements, typical investigation and numerical simulation shall be carried out for verification if necessary.

4.1.2 For the field investigation, the points and sections shall be rationally arranged according to the environmental status and change of the river and the region where the project is located, considering the relevance and coordination between the key elements such as the hydrological regime, water temperature, surface water environment, and aquatic ecosystem. The points and sections shall be representative, controlling and consistent.

4.2 Hydrological Regime

4.2.1 The hydrologic regime investigation shall meet the following requirements:

1. Collect the observation data on hydrological elements, including water level, flow rate, flow velocity and sediment, at the control sections in the reservoir area, downstream of the dam and in important tributaries.

2. Collect the data on water resource utilization of the project-affected river reach, including the domestic water, production water and ecological water, with consideration of the project purposes.

3. In the case that the project involves a sensitive water area with important ecological function or any other multiple-purpose utilization function, control sections shall be arranged to investigate the requirements of the sensitive area for hydrological elements such as water level, flow rate, flow velocity and water depth.

4. In the case that the inflow condition changes due to river hydropower development, regional water resource allocation, environmental function requirements and other external changes, the factors contributing to the inflow change shall be investigated. If necessary, runoff restoration computation shall be carried out. The runoff restoration computation shall comply with the current sector standard DL/T 5431, *Specification for Hydrologic Computation of Hydropower*

and Water Resources.

5 In the case that the observation data of the basin hydrological stations and other hydrological observation data cannot meet the requirements for the assessment, necessary hydrological observation shall be carried out.

4.2.2 The analysis and evaluation of the hydrological regime impact shall meet the following requirements:

1 Select proper sections for comparative analysis and evaluation accodording to the project characteristics and operation as well as the distribution of environmentally sensitive objects.

2 Analyze the temporal and spatial variations of water level in the reservoir and the downstream river channel according to the project operation characteristics.

3 Analyze the inter-annual variation, intra-annual variation and typical daily variation of the reservoir inflow and outflow on different time scales based on the reservoir inflows and outflows.

4 Select the typical flood hydrograph to analyze and evaluate the flood control functions of the reservoir such as flood storage and peak-shaving.

5 Analyze the temporal and spatial variations of the reservoir sedimentation and the sediment scouring and deposition in the downstream channel as well as their impacts, based on the inflow and outflow sediment concentrations, taking into account the project operation characteristics.

6 Analyze the variations of the hydrological elements on different time scales and their impacts according to the demands of the environmentally sensitive objects.

7 In the case that the project operates jointly with its upstream and downstream cascades, the role of the project in the joint operation and its actual impacts on the downstream hydrological regime shall be analyzed and evaluated on the basis of the analysis and evaluation on the actual impacts of the joint operation on the downstream hydrological regime.

4.2.3 Based on the results of the hydrologic regime investigation, analysis and evaluation, the representative parameters of the control sections shall be selected to analyze their consistency with the predicted results of the hydrological regime impact assessment, and the corrections to the hydrological regime impact prediction shall be put forward.

4.2.4 For the hydrological regime impact assessment, approaches such as hydrological statistics and comparative analysis should be used.

4.3 Water Temperature

4.3.1 The investigation on water temperature shall meet the following requirements:

1. Collect the observation data on the river water temperature of the project-affected area.

2. Collect the observation data on the meteorological elements from the representative meteorological observation station in the region, such as air temperature, irradiation, cloud cover, wind speed, humidity, and evaporation.

3. Collect the research data on the relationship between the river water temperature and the air temperature in the region.

4. Collect the data on the reservoir inflow water temperature, the vertical distribution of reservoir water temperature, the water temperature at dam toe, and the water temperature distribution along the river channel.

5. When there exists any environmentally sensitive object with special needs for water temperature in the project-affected water areas, the characteristics of the water-temperature-sensitive object and its requirements for water temperature shall be investigated.

6. When a stratified reservoir is constructed upstream and its operation results in variation in the reservoir inflow water temperature, the characteristics, construction time and operation mode of the upstream project, as well as the manner and extent to which the water temperature is affected by the upstream stratified reservoir shall be investigated.

7. When the observation data and research results of the water temperature in the river basin cannot meet the requirements of the post-assessment, supplementary observation on the water temperature shall be carried out. The supplementary observation on the water temperature may be conducted in accordance with the current sector standard NB/T 35094, *Code for Water Temperature Calculation of Hydropower Projects*.

4.3.2 The analysis and evaluation of the water temperature impact shall meet the following requirements:

1 Select proper sections to carry out comparative analysis and evaluation according to the project characteristics, water temperature investigation and distribution of the environmentally sensitive objects.

2 Analyze and evaluate the temporal and spatial variations of the river water temperature and its causes according to the cascade operation of the river and the regional climate change.

3 Analyze, with consideration of the project characteristics and operation, the variation pattern and influence factors of the water temperature in the reservoir area and downstream river channel before and after the construction of the project, including the characteristics of the water temperature structure, the vertical, horizontal and longitudinal distributions of the water temperature in the reservoir on different time scales, the outflow water temperature and its variation along the river channel.

4 Analyze the water temperature variation and its impacts according to the demands of the environmentally sensitive objects.

5 When the reservoir is of completely mixed type and the water temperature is nonstratified, the related water temperature impact evaluation should be adjusted according to the distribution and construction of the upstream stratified reservoirs.

6 When the operation of the upstream stratified reservoir affects the reservoir inflow water temperature, the actual impacts shall be analyzed and evaluated according to the impacts of the joint cascade operation on the hydrological regime, the construction time and operation mode of the upstream cascade, and the manner and extent to which the water temperature is affected.

4.3.3 The verification of the water temperature impact prediction shall meet the following requirements:

1 Analyze the consistency of the prediction results of the water temperature impact evaluation with the results of the water temperature investigation and analysis and evaluation.

2 Analyze the patterns, approaches and boundary conditions of water temperature impact, according to the actual impacts of the project operation on the water temperature, taking into account the variations of the inflow water temperature and the regional meteorological elements.

3 Compare the boundary conditions between the predicted and actual

water temperature impact, and analyze their differences and causes according to the prediction approach and model.

4   Recommend corrections to the prediction on the reservoir water temperature structure and the variation of the downstream water temperature along the river channel.

4.3.4 For the water temperature impact assessment, the statistics method and the comparative analysis method should be adopted.

4.4 Surface Water Environment

4.4.1 The surface water environmental investigation shall meet the following requirements:

1   Investigate the regional water environmental functions, and collect the monitoring data on surface water quality, substrate, eutrophication and total dissolved gas supersaturation at the related monitoring sections.

2   Investigate the types of main pollution sources in the reservoir area and the rivers flowing into the reservoir, pollution discharges, main pollutants, and operation of treatment facilities.

3   When there exists an environmentally sensitive object in the project-affected water area which has special demands on the water quality, the characteristics of the water quality-sensitive objects and their requirements for water quality shall be investigated.

4   When necessary, supplementary monitoring for the water quality shall be performed. The supplementary monitoring for the water quality shall comply with the current sector standard HJ/T 91, *Technical Specifications Requirements for Monitoring of Surface Water and Waste Water*.

4.4.2 Analysis and evaluation on the surface water impacts shall meet the following requirements:

1   Select suitable sections to carry out comparative analysis and evaluation according to the characteristic parameters of the pollution sources flowing into the reservoir and the variation of the reservoir hydrodynamic condition.

2   Analyze the surface-water quality, substrate status and changing trend of the project-affected water areas, and evaluate the satisfaction to the water environmental function objectives.

3   Analyze the variation of the surface-water quality and its impacts according to the demands of the environmentally sensitive objects.

4   Analyze nutrition status, the temporal and spatial distribution and causes of eutrophication of the reservoir water, with consideration of the analysis and evaluation results of hydrological regime and water temperature.

5   Analyze and evaluate the total dissolved gas supersaturation occurrence due to the project discharge water and the impacted scope and degree according to the investigation and monitoring data of the supersaturation of the river total dissolved gas.

4.4.3 The verification of the surface water impact prediction shall meet the following requirements:

1   Analyze the consistency of the prediction results of the surface water impact assessment with the results of the surface-water investigation, analysis and evaluation.

2   Analyze the pattern, approach and boundary conditions of the surface-water quality impacts according to the actual impacts of the project operation on the surface-water quality, taking into account the variation in the regional pollution sources.

3   Compare the boundary conditions between predicted and actual impacts, and analyze their difference and causes according to the prediction methods and models for the pollution sources, surface-water quality and water nutrition status.

4   Recommend corrections to the method and model for predicting the surface-water quality and eutrophication.

4.4.4   For the impact assessment of surface water, methods such as standard index method, comprehensive nutrition status index method and comparative analysis method should be used.

4.5   Aquatic Ecosystem

4.5.1   The investigation on the aquatic ecosystem shall meet the following requirements:

1   Investigate and collect the data on the investigation and evaluation results of aquatic ecosystem and fishery status in the river basin where the project is located, with emphases on the conservation functions, range, function zoning and main protected objects in the aquatic ecology-sensitive areas.

2   Investigate the structure and functions of the aquatic ecosystem in the project-affected water areas.

3   Investigate the species, biomass and distribution of phytoplankton, attached algae, zooplankton, benthos and aquatic vascular plants in the project-affected water areas.

4   Investigate the fish species, catches, fish biology, main fish habitats, fishes of early life history stage, and fishery status in the project-affected water areas, with emphases on the species, quantities and distribution of rare and endangered fishes, endemic fishes and important economic fishes, the fish migration routes, and the distribution and habitat features of spawning grounds, feeding grounds, and wintering grounds.

5   The methods for the aquatic ecosystem investigation shall comply with the current sector standard NB/T 10079, *Technical Code for Investigation and Assessment of Aquatic Ecosystem for Hydropower Projects*.

4.5.2 The analysis and evaluation on the aquatic-ecosystem impacts shall meet the following requirements:

1   Analyze the status and service functions of the aquatic ecosystem in the river basin according to the temporal and spatial variations of the species, quantity and distribution of aquatic organism in the river basin, especially the variations of the status of the important fish habitats such as the fish migration routes and the spawning grounds, feeding grounds, wintering grounds.

2   When the connectivity condition between the upstream and downstream aquatic ecosystems varies, the factors contributing to the aquatic ecosystem variation shall be investigated, and the pattern and degree of the impacts on the aquatic ecosystem due to the river development in the project-affected river reach shall be analyzed.

3   Analyze the actual impacts of the project construction and operation on the aquatic ecosystems and the species, quantity and distribution of the aquatic organisms in the reservoir area and the water area at dam toe, taking into account the variations of the water regime, water temperature and surface water environment in the project-affected water areas.

4   When aquatic ecology-sensitive areas are distributed in the project-affected water areas, the actual impact range and degree of the project construction and operation on the structure, functions and main protected objects in the aquatic ecology-sensitive areas shall be

analyzed and evaluated.

5   When important fish habitats such as spawning grounds, feeding grounds and wintering grounds are distributed in the assessment range, the impacts of the project on the important fish habitats and fish communities shall be analyzed and evaluated according to the requirements of the fishes for the water level, flow rate, flow velocity, water temperature, and water quality.

6   When the project is located at an estuary or the project operation affects the water regime, water temperature and surface water environment in the estuary area, the impacts of the project on the estuary aquatic ecosystem and the species, quantity and distribution of the aquatic organisms shall be analyzed and evaluated.

4.5.3 The verification of the aquatic-ecosystem impact prediction shall meet the following requirements:

1   Analyze the consistency of the prediction results of the aquatic ecosystem assessment with the results of the aquatic-ecosystem investigation, analysis and evaluation.

2   Analyze the pattern, process and boundary conditions of the aquatic-ecosystem impacts according to the actual impacts of the project construction and operation on the aquatic ecosystem, taking into account the variations in the water regime, water temperature and surface water environment.

3   Compare the predicted impacts of the aquatic ecosystem with the actual impacts in terms of pattern, process and boundary conditions, and analyze their difference and causes.

4   Recommend corrections to the aquatic ecosystem prediction.

4.5.4 For the aquatic-ecosystem impact assessment, approaches such as ecological mechanism analysis, accumulated impact evaluation, systematic analysis and comparative analysis should be used.

4.6   Terrestrial Ecosystem

4.6.1 The terrestrial ecosystem investigation shall meet the following requirements:

1   Investigate and collect the data on the investigation and evaluation results of terrestrial ecosystem and the forestry status in the region where the project is located, with emphases on the protection function, range, function zoning and main protected objects of the terrestrial

ecosystem-sensitive area.

2   Investigate the structure and function of the terrestrial ecosystem in the project-affected areas.

3   Investigate the plant species and flora composition, vegetation types and distribution as well as biomass status in the project-affected areas, with emphases on the species, quantity, distribution, growth status, habitat status and conservation status of the protected plants and narrow endemic plants as well as the species, quantity, ages, distribution, growth status, habitat status and conservation status of the old and notable trees.

4   Investigate the species, distribution, fauna, ecological habits, habitat conditions, population and structure of the terrestrial animals in the project-affected areas, with emphases on the species, distribution range, quantity, ecological habits, habitat distribution and conservation status of the protected animals.

5   The methods for terrestrial ecosystem investigation shall comply with the current sector standard NB/T 10080, *Technical Code for Investigation and Assessment of Terrestrial Ecosystem for Hydropower Projects*.

4.6.2   The analysis and evaluation on the terrestrial ecosystem impacts shall meet the following requirements:

1   Analyze the variations in vegetation types, vegetation coverage rate and plant species composition in the project-affected areas, with emphases on the impacts of the project on the protected plants, narrow endemic plants as well as old and notable trees.

2   Analyze the variations in the species, quantity, distribution and habitats of wild animals in the project-affected areas, with an emphasis on the impacts of the project on the protected animals.

3   Analyze the impacts of the project construction and operation on the regional biological diversity and landscape pattern according to the requirements of the regional ecological environment.

4   Analyze the actual impact range and extent of the project construction and operation on the quality of the regional ecological environment, the land use pattern and the landscape structure.

5   When any terrestrial ecology-sensitive area is located in the project-affected area, the actual impact range and extent of the project

construction and operation on the structure, functions and main protected objects in the ecology-sensitive area shall be analyzed and evaluated.

4.6.3 The verification of the terrestrial ecology impact prediction shall meet the following requirements:

1 Analyze the consistency of the prediction results of the terrestrial ecology assessment with the results of the terrestrial ecology investigation, analysis and evaluation.

2 Analyze the pattern, process and boundary conditions of the terrestrial ecology impacts according to the actual impacts of the project construction and operation on the terrestrial ecosystem, taking into account the climatic change.

3 Compare the predicted impacts of the terrestrial ecosystem with the actual impacts in terms of the pattern, process and boundary conditions, and analyze their differences and causes.

4 Recommend corrections to the terrestrial ecological prediction.

4.6.4 For the terrestrial ecology impact assessment, methods such as map overlays, ecological mechanism analysis and comparative analysis should be used.

4.7 Climate

4.7.1 The climate investigation shall meet the following requirements:

1 Collect the data on the basic climatic characteristics such as the general circulation of the atmosphere and the regional atmospheric circulation, the main disastrous weather, and the extreme climate in the region where the project is located.

2 Collect the observation data on meteorological elements, such as air temperature, precipitation, evaporation, humidity, wind speed, wind direction, air pressure, frost-free period, and foggy days, from the meteorological stations in the region where the project is located, and analyze the representativeness and accuracy of the data.

3 When snowy mountain, glacier and rime are distributed in the project-affected areas, their types, distribution, scale, seasonal variation and key meteorological factors shall be investigated.

4.7.2 Climatic impact analysis and evaluation shall meet the following requirements:

1   Analyze the relationship of the climate change trend in the region where the project is located with the global climate change trend.

2   Select the representative and reference meteorological stations, and analyze the temporal and spatial variation characteristics of meteorological elements such as air temperature, precipitation, evaporation, humidity, wind speed, wind direction, frost-free period, and foggy days, as well as the variation characteristics of air temperature, wind speed, wind direction, relative humidity and barometric pressure at different gradients, with an emphasis on the change of meteorological elements in climate-sensitive periods.

3   Analyze and evaluate the range and extent of the impact of the project on the regional climate and the reservoir-based local climate with consideration of the project operation pattern, and identify whether the hydropower project construction has led to any disastrous climate such as drought, rainstorm, high temperature, low temperature, and rainy and/or snowy days.

4.7.3   The verification of climate impact prediction shall meet the following requirements:

1   Analyze the consistency of the climate impact prediction results with the outcomes of the climate investigation, analysis and evaluation.

2   Analyze the pattern, process and boundary conditions of the project impact on the regional climate and reservoir-based local climate according to the project characteristics and operation.

3   Compare the predicted and actual climatic impacts in terms of the pattern, process and boundary conditions, and analyze their difference and causes.

4   Recommend corrections to the approach and model of the climatic impact prediction.

4.7.4   For the climate impact assessment, comparative analysis method and statistical models should be adopted.

4.8 Socio-economy

4.8.1   The socio-economy investigation shall meet the following requirements:

1   Investigate the land use status and planning in the region where the project is located, with emphases on the amount and distribution of farmland resources, amount and distribution of ecologically protected land, land planned for urban development and important

infrastructure, etc.

2 Investigate the total water resource amount and water resources planning and utilization status in the region where the project is located, and analyze the water use demands in the project-affected water areas and important water withdrawal facilities including the withdrawal amount, water level, period, etc.

3 Investigate the economic aggregate, industrial structure, fiscal revenue, income of residents, employment status, and the cultural, educational, health and transportation facilities in the region where the project is located.

4 Investigate the achievement of project purposes such as power generation, flood control, navigation, irrigation, ecological protection, and tourism.

4.8.2 The socio-economic impact analysis and evaluation shall meet the following requirements:

1 Analyze the variation in types, quantity and utilization patterns of land resource, analyze and evaluate the variation in water resource development and utilization, and evaluate the impacts of the project construction and operation on the land and water resource.

2 Investigate and analyze the variation in economic aggregate, industrial structure, fiscal revenue, income of residents, employment status, and cultural, educational, health and transportation facilities in the region where the project is located, and evaluate the impacts of the project construction and operation on the regional socio-economy.

3 Analyze and evaluate the project benefits such as power generation, flood control, navigation, irrigation, ecological protection, tourism development, energy conservation and emission reduction, energy mix optimization, water resources utilization, and fishery aquaculture.

4.8.3 For the socio-economic impact assessment, methods such as mathematical statistics method, input-output model and comparative analysis should be used.

4.9 Environmental Impacts of Resettlement

4.9.1 Resettlement investigation shall meet the following requirements:

1 Investigate the resettlement population, pattern, land development and utilization, infrastructure construction, environmental protection facility construction, and ecological protection.

2 Investigate the variations of the ecological environment in the resettlement areas.

4.9.2 The analysis and evaluation on the resettlement impacts shall meet the following requirements:

1 Analyze the actually resettled people, the number and distribution of resettlement sites, and the variations in the production and living conditions, income and livelihood of the resettled people, and ethnic culture.

2 Analyze and evaluate the impacts of the resettlement on the regional ecological environment according to the variations in the ecological environment of the resettlement area, taking into account the construction of the environmental protection facilities and ecological protection works.

4.9.3 For resettlement environmental impact assessment, mathematical statistics and comparative analysis methods should be adopted.

4.10 Comprehensive Evaluation

4.10.1 The investigation, analysis and evaluation on environmental impacts shall, on the basis of environmental impact investigation, systematically analyze the environmental quality status, the ecological environment status, socio-economic status, main environmental problems and their causes in the project-affected areas.

4.10.2 The investigation, analysis and evaluation on environmental impacts shall, according to the results of the environmental impact analysis and evaluation, comprehensively analyze the pattern and degree of the impact of the project construction and operation on the environmental quality, ecological environment and socio-economy in the region.

4.10.3 The investigation, analysis and evaluation on environmental impacts shall, according to the pattern and extent of the impacts of the project operation on the regional environment, analyze the environmental change trend in the project-affected areas, taking into account the ecological environmental protection, water resource development and protection, river development and utilization.

5 Investigation and Assessment of Environmental Protection Measures

5.1 General Requirements

5.1.1 The design, construction and operation of environmental protection measures shall be investigated according to relevant provisions in the river development plan, the project environmental impact assessment documents, the project design documents and other related documents. The conformity with the design of environmental protection measures shall be analyzed.

5.1.2 The effectiveness of the environmental protection measures shall be evaluated according to the operation of the measures, taking into account the environmental quality and the status and changing trend of the ecological environment in the project-affected areas; the existing problems shall be analyzed; the comments and recommendations on the improvement shall be put forward from the perspective of the whole river basin.

5.2 Hydrological Regime

5.2.1 For the hydrological regime mitigation measures, the design, construction and operation of the ecological discharge facilities and the design, implementation and management of the ecological operation scheme shall be investigated; the achievement extent of the ecological discharge measures shall be analyzed in terms of the reservoir water level, and the scale, duration and dependability shall be analyzed, with an emphasis on the extent to which the design objectives are achieved.

5.2.2 The evaluation on the effectiveness of the ecological discharge measures shall, according to the regional water use demands, analyze the rationality of the ecological flow and assess the environmental protection effectiveness of ecological discharge measures and ecological operation.

5.2.3 The comments and recommendations on improving the water regime mitigation measures shall be put forward through analysis of the existing problems on the basis of the evaluation on the effectiveness of the ecological discharge measures and the ecological operation, taking into account the requirements of the environmentally sensitive objects.

5.3 Water Temperature

5.3.1 For the water temperature mitigation measures, the design, construction and operation of the water temperature mitigation facilities and monitoring system shall be investigated, with emphases on the selective water intaking facility and the ecological operation scheme. Against the construction items in the design scheme and implementation scheme of the water temperature mitigation measures, the design parameters, and the design objectives, the achievement extent of the water temperature mitigation measures shall be analyzed, with an emphasis on the extent to which the design objectives are achieved.

5.3.2 For the assessment on the effectiveness of the water temperature mitigation measures, the mitigation of the water temperature impacts shall be analyzed and evaluated according to the water temperature monitoring results and operation condition. With consideration of the aquatic ecosystem investigation, the impact analysis and evaluation and the impact degree of the environmentally sensitive objects, the effectiveness of the water temperature mitigation measures shall be analyzed and evaluated in terms of design, operation, management, etc.

5.3.3 The comments and recommendations for improving water temperature impact mitigation measures shall be put forward on the basis of analyzing the existing problems in the mitigation measures of the water temperature impact and the causes and with consideration of the protection requirements of the environmentally sensitive objects on water temperature.

5.4 Surface Water Environment

5.4.1 For the surface water protection measures, the design, construction and operation of the measures such as engineering works, risk control, operation, water quality monitoring and management adopted for protecting the water quality of the reservoir area and the downstream area shall be investigated, with emphases on the control measures for the reservoir pollution sources around the reservoir, the operation scheme for safeguarding the downstream water quality and the water quality monitoring system. Against the contents in the design scheme and implementation scheme of the surface-water protection measures, the design parameters, and the design objectives, the extent to which the surface water protection measures are achieved shall be analyzed.

5.4.2 For the assessment on the effectiveness of the surface water protection measures, the implementation effectiveness shall be analyzed in terms of the water quality indicators, nutrition status and pollutant bearing capacity. The satisfaction degree of the water pollution control objectives, maintenance of the water environment functions, water environment improvement, etc. in the project-affected water areas shall be evaluated.

5.4.3 The comments and recommendations on improving the surface water protection measures shall be put forward with respect to the existing problems in the surface water protection measures and the causes, taking into account the requirements of the water environment function.

5.5 Aquatic Ecosystem

5.5.1 For the aquatic ecosystem protection measures, the design, construction and operation of the aquatic ecosystem protection measures shall be investigated, with emphases on the ecological operation scheme, habitat protection measures, fish passage facilities, and fish restocking measures. Against the contents in the design scheme and the implementation scheme of the aquatic ecosystem protection measures, the design parameters, and the design objectives, the implementation extent of the aquatic ecosystem measures shall be analyzed.

5.5.2 For the evaluation on the effectiveness of the habitat protection, the fish habitat protection scheme, facility type and structure effectiveness shall be investigated in terms of river connectivity and the habitat protection, and the effectiveness of the fish habitat protection measures shall be analyzed and evaluated.

5.5.3 For the assessment on the effectiveness of fish passing, the effectiveness of the fish passage facilities shall be analyzed and evaluated according to the operation of the fish passage facilities and the auxiliary facilities.

5.5.4 For the assessment on the effectiveness of the fish restocking, the fish restocking shall be investigated, with emphases on the scale, objects, production process, raising mode, scientific research, etc. of the fish restocking stations. The effectiveness of the fish restocking measures shall be analyzed and evaluated.

5.5.5 The comments and recommendations for improving the aquatic ecosystem protection measures shall be put forward with respect to the existing problems in the aquatic ecosystem protection measures and the causes, taking into account the requirements for the river basin aquatic ecosystem protection.

5.6 Terrestrial Ecosystem

5.6.1 For the assessment on the terrestrial ecosystem protection measures, the design, construction and operation of the terrestrial ecosystem protection measures such as avoidance, mitigation, compensation and repair adopted for the project shall be investigated, with emphases on the protection measures for the protected plants, narrow endemic plants and old and notable trees. Against the construction items in the design scheme and the implementation scheme of the terrestrial ecosystem protection measures, the design parameters, and the design objectives, the achievement extent of the terrestrial ecosystem protection measures shall be analyzed.

5.6.2 For the assessment on the effectiveness of the terrestrial ecosystem protection measures, the implementation effectiveness of the terrestrial ecosystem protection measures shall be analyzed and evaluated according to the variations in vegetation classification, plant species and quantity and regional ecological stability in the project-affected areas, with emphases on the protection achievements of the protected plants, narrow endemic plants and old and notable trees.

5.6.3 The comments and recommendations for improving the terrestrial ecosystem protection measures shall be put forward with respect to the existing problems in the terrestrial ecosystem protection measures and the causes, taking into account the requirements on the river basin terrestrial ecosystem protection.

5.7 Environmental Protection of Resettlement

5.7.1 For the resettlement environmental protection measures, the implementation of measures for restoration of the production and living conditions, water supply, landscape and cultural relics protection, ethnic culture preservation, etc. and the construction of environmental protection facilities in the resettlement area shall be investigated.

5.7.2 For the evaluation on effectiveness of the resettlement environmental protection measures, the satisfaction degree of resettlement on economic and social development, environmental quality management and ecological functions shall be evaluated on the basis of effectiveness analysis of the measures.

5.7.3 The comments and recommendations on improvement of the resettlement environmental protection shall be put forward with respect to the existing problems in the resettlement environmental protection.

5.8 Comprehensive Evaluation

5.8.1 The comprehensive evaluation of the environmental protection measures shall, on the basis of investigation and analysis, comprehensively analyze the systematicness and effectiveness of the environmental protection measures implemented for the project.

5.8.2 The comprehensive evaluation of the environmental protection measures shall, from the systematic perspective, put forward the improvement comments and recommendations according to the quality status and changing trend of the regional environment, existing problems and policy requirements for the project environment protection.

6 Survey of Public Opinions

6.1 General Requirements

6.1.1 The survey of public opinions shall cover the opinions on the project environmental protection work of the community public, stakeholders and project-affected groups.

6.1.2 For classified items, the relevant national regulations shall be complied with.

6.2 Survey Methods and Respondents

6.2.1 The methods for the survey of public opinions include field interview and workshop, expert consultation, media reporting, academic research and analysis, complaints analysis, public opinion analysis, questionnaire, etc.

6.2.2 The respondents shall be selected according to the methods and contents of the survey. The respondents for the public opinion shall be extensive and representative, consisting of the public in the project-affected area, the public concerning the project construction, the related competent authorities, experts, etc., with emphases on the stakeholders and project-affected groups.

6.3 Survey Contents

6.3.1 The range of the survey of public opinions should be focused on the project-affected area. The survey range may be properly extended according to the requirements of the assessment elements on the premise of satisfying the representativeness.

6.3.2 The contents of the survey may be set up according to the characteristics of the project and its surroundings, including the following:

1. Whether any events of environmental pollution or ecological damage have ever occurred since the project construction.

2. Public perception and understanding of the main environmental problems existing in the construction and operation of the project and the environmental impacts.

3. Public satisfaction with the outcomes of the environmental protection measures adopted in the project construction and operation periods and other opinions.

4. Environmental problems highly concerned by the public and recommendations on further environmental protection measures.

5 The general public evaluation on the project environmental protection work.

6.4 Analysis and Addressing of Public Opinions

6.4.1 The analysis on the survey data shall meet the following requirements:

1 Conduct categorized statistics on the public opinions and put forward the corresponding solutions.

2 State the public acceptance degree on the project environmental protection work and the main public opinions on the project.

3 State the solutions to the main public opinions and feedback.

4 Take the main environmental problems which attract public concerns and strong response as the important references for presenting comments and recommendations on subsequent improvement measures.

6.4.2 The addressing of the public opinions shall analyze the results of the public opinion statistics and put forward the solutions to the relevant problems.

7 Conclusions and Recommendations

7.0.1 Post-assessment of environmental impacts for a hydropower project shall systematically retrospect the design, construction and operation of the project and analyze the environmental rationality of the project changes.

7.0.2 Post-assessment of environmental impacts for a hydropower project shall comprehensively analyze the changing trend of the regional environmental quality and the causes according to the project operation, taking into account the regional socio-economic development.

7.0.3 Post-assessment of environmental impacts for a hydropower project shall summarize the actual impacts arising from the project operation, and analyze the difference between the actual and predicted environmental impacts, as well as the causes.

7.0.4 Post-assessment of environmental impacts for a hydropower project shall summarize the implementation and operation of the environmental protection measures, analyze their reliability, adaptability and effectiveness as well as the ecological protection effect, and put forward practical comments and recommendations on improvement.

7.0.5 For long-term and cumulative environmental impacts that have not appeared in the short project operation period, recommendations on follow-up investigation, monitoring, monographic research, etc. shall be put forward.

7.0.6 Post-assessment of environmental impacts for a hydropower project shall put forward recommendations on the adaptive management for the hydropower development according to the general requirements for ecological and environmental management of the river basin.

Appendix A Table of Contents for a Report on Post-assessment of Environmental Impacts of Hydropower Projects

Foreword

1 General
1.1 Background
1.2 Assessment Purpose
1.3 Assessment Basis
1.4 Assessment Range
1.5 Assessment Content
1.6 Assessment Indicators
1.7 Assessment Period
1.8 Environmental Protection Objectives
1.9 Assessment Approach

2 Project Review
2.1 River Hydropower Planning, Development and Utilization
2.2 Project Construction and Operation
2.3 Environmental Rationality Analysis of Project Change
2.4 Environmental Protection Work

3 Analysis and Assessment on Regional Environmental Changing Trend
3.1 Environmental Quality
3.2 Ecological Environment
3.3 Socio-economy

4 Analysis and Verification of Project Environmental Impacts
4.1 Hydrological Regime
4.2 Water Temperature
4.3 Surface Water
4.4 Aquatic Ecosystem

4.5	Terrestrial Ecosystem
4.6	Climate
4.7	Socio-economy
4.8	Environmental Impacts of Resettlement
4.9	Comprehensive Assessment
5	**Investigation and Assessment on Environmental Protection Measures**
5.1	Hydrological Regime
5.2	Water Temperature
5.3	Surface Water
5.4	Aquatic Ecosystem
5.5	Terrestrial Ecosystem
5.6	Environmental Protection of Resettlement
5.7	Comprehensive Assessment
6	**Survey of Public Opinions**
6.1	Survey Methods and Respondents
6.2	Survey Contents
6.3	Analysis and Addressing of Public Opinions
7	**Conclusions and Recommendations**
7.1	Conclusions
7.2	Recommendations

Attachments:

1	Approval of River Hydropower Development Plan
2	Examination Comments on Environmental Impact Assessment Documents in River Hydropower Planning Stage
3	Approval of Environmental Impact Assessment Documents
4	Examination Comments on the Feasibility Study Report
5	Others

Attached drawings:

1	Geographic Location Map

2	River Hydropower Development Sketch
3	Hydrographic Map
4	Project General Layout
5	Reservoir Inundation Map
6	Distribution and Variation Map of Environmentally Sensitive Objects
7	Distribution and Variation Map of Important Fishes
8	Distribution and Variation Map of Main Fish Habitats
9	Distribution and Variation Map of Vegetation Types
10	Distribution and Variation Map of Land Utilization Types
11	Distribution Map of Key Protected Wild Plants and Animals and Old and Notable Trees
12	Environmental Protection Measures Layout
13	Others

Appendix B Indicators for Post-assessment of Environmental Impacts for Hydropower Projects

Table B Indicators for post-assessment of environmental impacts for hydropower projects

Environmental element	Assessment indicator	Assessment criteria	Assessment method
Water regime	Water level (m) and flow rate (m³/s) of control sections and their temporal variations	Satisfaction degree to ecological water demand	Comparative analysis
	Water level (m), flow rate (m³/s), flow velocity (m/s), and water depth (m) of sensitive water areas and their temporal variations		Comparative analysis
	Flood hydrograph and its temporal and spatial variations (m³/s, %)		Comparative analysis
	River channel sediment transport (t)	—	Comparative analysis
Water temperature	Reservoir water temperature structure	—	Comparative analysis
	Outflow water temperature and its variation along the river channel (°C, %)	Ecological threshold	Comparative analysis
Surface water	Water qualification rate (%)	Water quality criteria for the water environment function areas	Standard index

31

Table B (*continued*)

Environmental element	Assessment indicator	Assessment criteria	Assessment method
Surface water	Eutrophication status of the reservoir water body	Lake tophic status grading standard	Comprehensive nutrition state index
	Supersaturation of total dissolved gas (%) and occurrence frequency (d/a)	Limit value acceptable to fishes	Comparative analysis
Aquatic ecosystem	Species composition, dominant groups and distribution of aquatic organisms		Ecological mechanism analysis, comparative analysis
	Fauna, population and species of fishes		Ecological mechanism analysis, comparative analysis
	Species, quantities and distribution of the protected fishes	Whether the ecosystem changes or species are extinguished due to the change	Ecological mechanism analysis, comparative analysis
	Distribution of main fish habitats		Map overlays, comparative analysis
	Changes in area, distribution, structure and functions of aquatic ecosystem-sensitive areas		Map overlays, comparative analysis
	Changes in status and service functions of the aquatic ecosystem		Ecological mechanism analysis, comparative analysis

Table B (*continued*)

Environmental element	Assessment indicator	Assessment criteria	Assessment method
Terrestrial ecosystem	Changes in vegetation types, succession regularities and coverage rates		Ecological mechanism analysis, comparative analysis
	Changes in species, quantity and distribution of protected plants, narrow endemic plants, and old and notable trees, and their growth status	Whether the ecosystem changes and species are extinguished due to the change	Comparative analysis
	Changes in species, faunas, distribution and populations of terrestrial animals and their growth status		Ecological mechanism analysis, comparative analysis
	Changes in area, distribution, structure and functions of terrestrial ecosystem-sensitive areas		Map overlays, comparative analysis
	Changes in status and service functions of the terrestrial ecosystem		Ecological mechanism analysis, comparative analysis
Climate	Air temperature (°C), precipitation (mm), evaporation (mm), humidity, wind speed (m/s), wind direction, and fog and their changes (%)	—	Comparative analysis
	Occurrence probability (d/a) of extreme and damaging weather and its changes (%)	—	Comparative analysis

Table B *(continued)*

Environmental element		Assessment indicator	Assessment criteria	Assessment method
Social environment	Economic development	Benefits from the purposes such as power generation, flood control, water supply, irrigation, and navigation	Development purposes and objectives	Mathematical statistics
		Contribution to GDP, and effect on industrial structure and infrastructure	—	Mathematical statistics, comparative analysis
	Land resources	Changes in types, amounts, use, etc. of land resources	—	Comparative analysis
	Water resources	Dependability of various water uses and their changes	—	Comparative analysis
	Resettlement	Changes in production and living conditions, income and livelihood of the relocatees	—	Comparative analysis
		Changes in resource and environment bearing capacity related to resettlement	—	Comparative analysis
		Change in ethnic culture	—	

Explanation of Wording in This Code

1. Words used for different degrees of strictness are explained as follows in order to mark the differences in executing the requirements in this code:

    1) Words denoting a very strict or mandatory requirement:

        "Must" is used for affirmation; "must not" for negation.

    2) Words denoting a strict requirement under normal conditions:

        "Shall" is used for affirmation; "shall not" for negation.

    3) Words denoting a permission of a slight choice or an indication of the most suitable choice when conditions permit:

        "Should" is used for affirmation; "should not" for negation.

    4) "May" is used to express the option available, sometimes with the conditional permit.

2. "Shall meet the requirements of... " or "shall comply with... " is used in this code to indicate that it is necessary to comply with the requirements stipulated in other relative standards and codes.

List of Quoted Standards

NB/T 10079, *Technical Code for Investigation and Assessment of Aquatic Ecosystem for Hydropower Projects*

NB/T 10080, *Technical Code for Investigation and Assessment of Terrestrial Ecosystem for Hydropower Projects*

NB/T 35094, *Code for Water Temperature Calculation of Hydropower Projects*

DL/T 5431, *Specification for Hydrologic Computation of Hydropower and Water Resources*

HJ/T 91, *Technical Specifications Requirements for Monitoring of Surface Water and Waste Water*